Structures and Functions of Living Things

Reader

Core Knowledge®

Printed in Canada

ISBN: 978-1-68380-526-7

Structures and Functions of Living Things

Table of Contents

Biological Organization

Chapter 1

Look closely at your hand. What do you see? You see skin. Now squeeze your fingers and wrist with your other hand. You feel bones and some soft parts, too. Bones and skin are made up of even smaller parts called cells. A **cell** is the smallest unit of life. Cells are the basic unit of all living things.

Some tiny organisms consist of just one cell. These are called unicellular. These organisms are so small that you cannot see them without a microscope.

Many organisms, such as plants and animals, are made of many cells. They are called multicellular. Each type of cell has its own job. All the cells work together to help keep an organism alive.

Big Question

How are cells and tissues related in living things?

Vocabulary

cell, n. the smallest unit of life

Word Parts

The prefix *uni-* means one.

The prefix *multi-* means many.

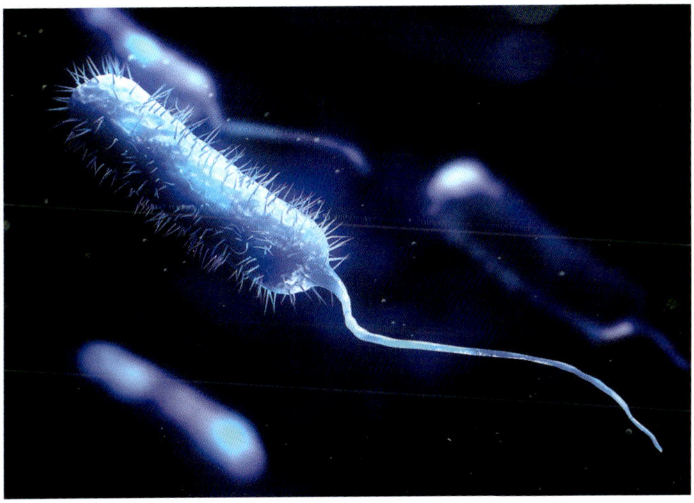

Bacteria are organisms that consist of only one cell. An individual is called a bacterium.

Cells Are the Building Blocks of All Animal Life

Animals are made of many kinds of cells. Each type of cell has its own role. Skin cells, blood cells, nerve cells, and muscle cells are just a few of the kinds of animal cells.

Skin cells form a layer. This layer covers the whole body. In this way, it protects all the parts inside an animal's body. Special skin cells make fur, hair, and nails.

Blood carries oxygen throughout an animal's body. Blood also contains cells that help to keep the animal from getting sick.

Nerve cells help to send messages to and from an animal's brain. They deliver these signals to the rest of the animal's body. Some nerve cells connect to muscles. Their signals tell the muscles to move. In turn, muscle cells help an animal to move. Special muscle cells cause the heart to beat.

These are a type of cell found in blood. These cells fight infection throughout an animal's body.

Cells Are the Building Blocks of All Plant Life

Like animal cells, types of plant cells have their own jobs that help to keep plants alive. Leaves are made of cells. Some of these cells make or store food for the plant. Other leaf cells allow the plant to take in air that it needs. Another type of leaf cells form tubes that carry food from the leaves to the rest of the plant.

Stems and roots are made of cells, too. Bundles of cells make the stem stiff and strong. Tubes in the stem, which develop from cells, allow water to move. Root cells take in water from the soil. The water moves from roots through the stem to the leaves.

This photo taken through a microscope shows cells in the leaf of a plant.

Groups of Similar Cells Form Tissues

Groups of similar kinds of cells form **tissues**. In animals, groups of muscle cells that work together make up muscle tissue. Muscle tissue allows an animal to move.

Vocabulary

tissue, n. a group of similar cells joined together

Bone tissues are made of bone cells. Bones protect soft parts inside an animal and help to add a rigid structure to give an animal shape. Connective tissues help hold different body parts together. Each of these types of tissues is made of a group of similar cells.

Plants have several tissues too. Some tissue forms the outside of a plant—its protective layer. Another kind of tissue transports water and nutrients to the plant's stems and leaves. Yet another type of tissue forms the growing parts of the plant at the tips of stems and roots. Each of these types of tissues is made of a group of similar cells.

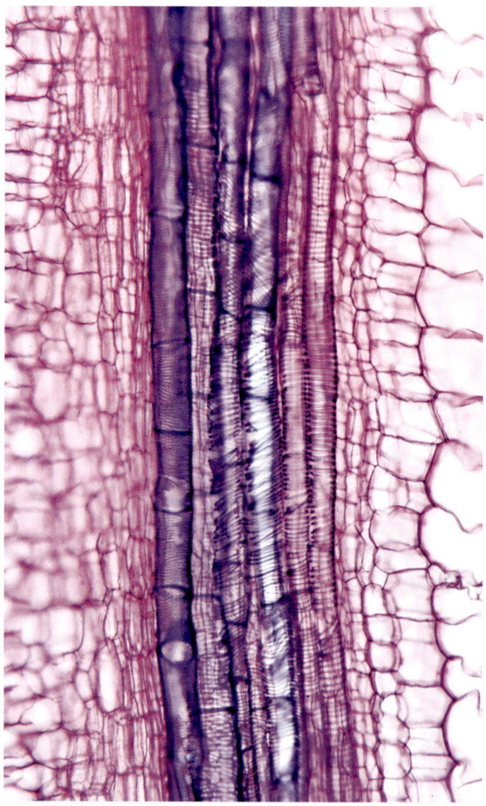

Tissues inside plants form tube-like structures. This type of tissue carries water from the roots, through the stem, and to the leaves.

In all living things, cells make up tissues. This pattern of structure is what scientists call *biological organization.*

Organs and Organ Systems

Chapter 2

You don't need to think about what goes on inside your body to do all the things you do every day. Your heart beats. Your lungs breathe. Your stomach digests food. Your intestines absorb nutrients. Your muscles stretch and move. All of the parts of your body have jobs. They all work together to keep you alive.

This all occurs because your body is highly organized. You are made of smaller parts that work together in a system. The parts, or **structures**, perform jobs, or **functions**.

Big Question

How do organs work together in systems to perform specific functions?

Vocabulary

structure, n. the arrangement of parts that make up something

function, n. the way that something works to achieve a task or serve a purpose

Both *structure* and *function* can also be used as verbs.

What parts of your body work together to enable you to ride a bike?

Groups of Tissues Form Organs

In the last chapter, you learned about cells and tissues. Groups of similar cells form a tissue.

Groups of similar tissues form body parts called organs. An **organ** is a structure in a living thing's body. Each type of organ has a structure and function. Animals have many organs. For example, the heart is an organ. A blue whale's heart is large, and a mouse's heart is very small. But in both animals, the heart's function is to pump blood through the animal's body. Other organs, each of which is formed from tissues, include the stomach, the liver, and the large intestine.

> **Vocabulary**
>
> **organ, n.** a body part made up of related tissues that perform a specific function

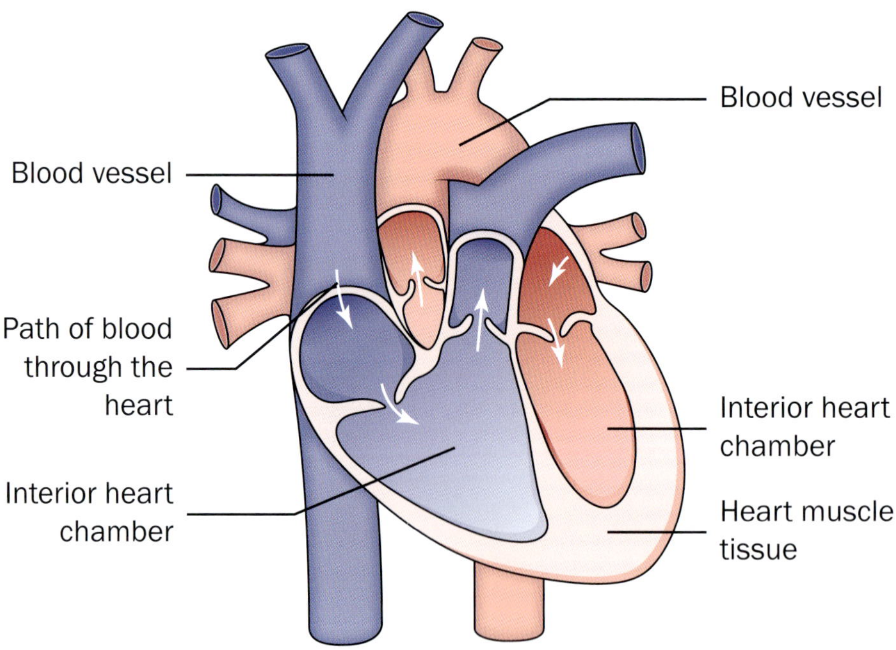

A heart is an organ, and it is made up of groups of tissues working together.

An Organ's Structure Supports Its Function

Animals that live on land have lungs. Lungs are expandable, air-filled organs. They allow an animal to breathe. Breathing takes in oxygen from the air. But lungs cannot function in water. Fish live in water and do not have lungs. They have organs called gills. Gills are blood-filled, sheetlike organs. A fish's gills allow it to get oxygen from the water instead of from air. Both lungs and gills are organs made of tissues.

Word to Know

Oxygen is a gas. Land-dwelling animals take in oxygen from the air. Aquatic organisms often take oxygen that is in the water through organs called gills.

Plants have organs, too. Leaves, stems, and roots are organs. The function of a plant's leaves is usually to capture sunlight to make food. A stem supports a plant and moves water to its different parts. Roots take in water from soil. A plant uses all of these organs and their functions to survive. And all of the plant's organs are made of tissues.

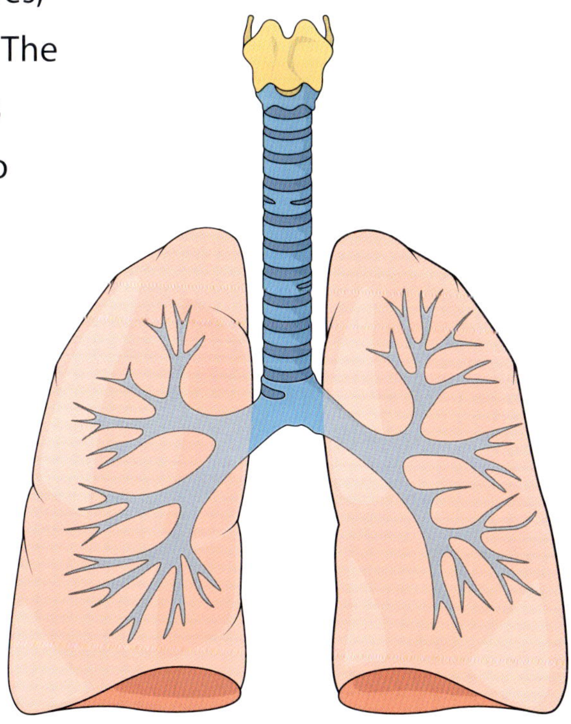

Lungs are part of the respiratory system. The respiratory system is what helps many organisms to breathe, or get air to different parts of their bodies.

Groups of Organs Form Organ Systems

Groups of similar cells form a tissue. Groups of tissues form organs. Groups of organs form **organ systems**. Organ systems have different structures, and each is related to the organ system's function.

> **Vocabulary**
>
> **organ system, n.** a group of organs that work together to carry out certain functions

Getting Food and Releasing Waste

Plants make their food by using energy from the sun. To do this and to process that food, they have organ systems. Three organs, the roots, stems, and leaves, all work together in a system that uses sunlight, carbon dioxide, and water to make food. Organs in the root system work together to help plants get water and store food. They use this food for all life processes.

Animals must find their own food. They use organs, their eyes and ears, to locate food. They use organs such as the stomach and intestines to digest the food they eat. This is called the digestive system. The kidney and bladder are organs that work together as a system to remove waste from the body. They are part of the excretory system, which excretes, or releases, waste.

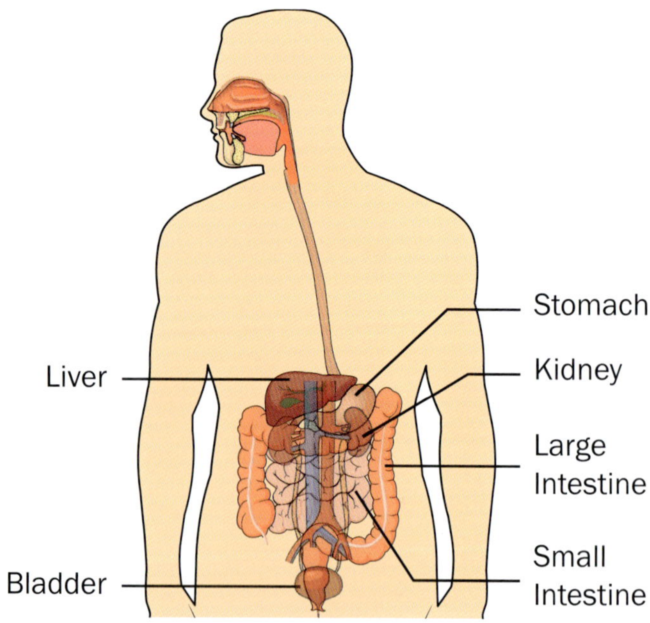

The digestive system is made up of organs that help animals to process the food they take in.

Exchanging Gases

When an animal breathes, it uses its lungs to bring in oxygen from the air. Muscles perform the pumping action that draws air into the lungs and pushes it out. Inside the lungs, gases cross to and from the animal's blood. Blood absorbs oxygen. That oxygen is used in the body to help make usable energy for life. Carbon dioxide is released in this process. It is released when the lungs exhale air from the body. This process is respiration, and the system is called the respiratory system.

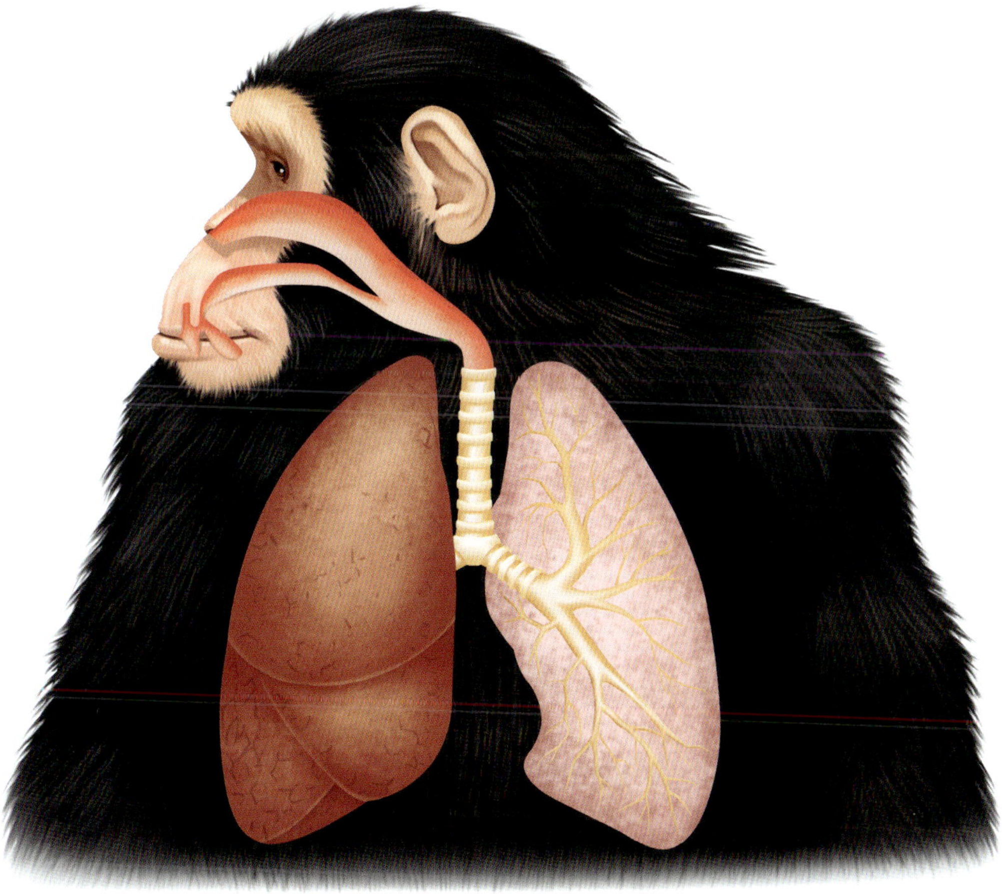

The respiration system is inside the chimpanzee's body. Can you trace the path that gases take as they enter and leave the chimpanzee's body?

Controlling Movement

Animals must move to meet their needs. They run, swim, jump, and climb. Organs of the muscular system work together so that animals can move.

Muscles are flexible. They can bend and stretch when an animal moves. Bones, which make up the skeletal system, are not flexible. They support and protect an animal's body. Muscles push and pull against stiff bones, causing movement. Animals would not be able to move without bones or muscles. These organ systems work together.

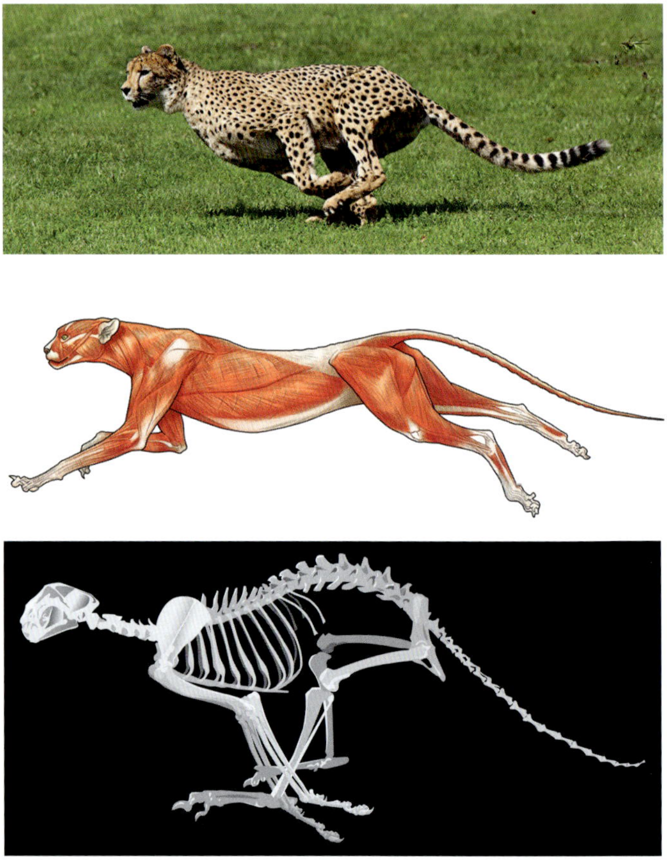

Cheetahs use many systems when running fast, including their muscular and skeletal systems.

Most muscles and all bones do not move on their own. The brain sends messages to these body parts through **nerves**. Nerves also send messages back to the brain. The brain and nerves are also part of an organ system. This is known as the nervous system.

Vocabulary

nerve, n. a bundle of cells that sends and receives messages between the brain and other body parts

Functions in Animals

Chapter 3

Cells, tissues, organs, and organ systems are levels of biological organization. The function of this organization is to help the organism to **survive**, or continue to live.

Animal structures support functions that are needed for **survival**, or staying alive. Different kinds of animals can have very different structures. But animals all have in common that their structures function to support their life processes.

Big Question

How do some animal structures function?

Vocabulary

survive, v. to stay alive

survival, n. the state of continuing to live

This koala is sleeping. It may not seem like the koala is doing anything, but sleep is one of many important body functions needed for survival.

Animal Structures Support Growth Functions

When a baby bird hatches from an egg, it needs help to survive. It cannot protect itself or find its own food. It needs to grow bigger and stronger so that it can stay alive. For **growth** to occur, the bird needs food, water, and air. These are needs that all animals have. Systems of organs in the animal's body help it

> **Vocabulary**
>
> **growth, n.** the life process of becoming bigger and stronger

- chew and digest food to get energy and the materials their bodies need to grow tissues and organs, such as muscles, bones, skin, and other structures;
- exchange gases with the surrounding air or water; and
- get rid of waste.

Young organisms sometimes struggle to survive on their own. While they are growing, young organisms' structures may not be developed enough to function fully to allow the organism to survive. They often have to rely on parents to help them meet their needs. These needs can include finding food, water, and shelter. Without these things, the young cannot grow into adults that will reproduce.

These baby birds need food, water, and air to grow into adults. Like these birds, many animals rely on their parents to help them survive when they are very young.

Animal Structures Support Behavioral Functions

Different kinds of animals act in different ways. The way an animal acts to survive is known as **behavior**. For example, spiders spin webs to help catch food. Horses graze to find their food. Animals use their structures to carry out these behaviors. Spiders have structures that create webbing. Horses have mouth and tooth structures that make it easier to nip and chew grasses.

Vocabulary

behavior, n. the ways in which an organism acts

When we see an animal move, it is performing a behavior. Animals have structures that enable their types of movement. A hawk's wings allow it to glide as it looks for food on the ground below. An octopus's soft body lets it squeeze into small spaces to hide.

Hunting and catching prey are also behaviors. Predators have structures that enable them to catch prey. A cheetah's tail helps it balance as it changes direction during a chase. A lion's sharp teeth and claws make it able to grab prey and tear meat.

Sharp teeth are structures that help a lion grab and hold prey.

Animal Structures Support Reproductive Functions

When animals survive long enough to grow into adults, they are able to reproduce. When animals **reproduce**, they make babies, or offspring. The offspring are new, young individuals of the same type as the parents. The ability to reproduce is a basic life process of all organisms.

Vocabulary

reproduce, v. to make new, young individuals of the same type

Individual animals do not have to reproduce to survive. However, when animals do reproduce, their offspring become the next generation. Any type of plant or animal survives only when some individuals succeed in reproducing.

Different animals reproduce in different ways, but most animals mate in pairs. Animals have different structures that aid in reproduction. Most mammals have structures to give birth to live young. All birds and most fish, reptiles, and amphibians produce eggs that hatch outside the mother's body.

Primates are mammals. They have structures for reproducing. The primates in the picture have additional structures that help the young to cling to the parents.

Functions in Plants

Chapter
4

In the last chapter, you learned that an animal's structures work together to perform functions. The organs and organ systems of animals support processes that are necessary for survival. Plants also have structures that work together to keep themselves alive. Types of plants can look very different from one another. But many plants share necessary structures in common, such as roots, stems, and leaves. These kinds of plant structures support life processes that ensure a plant grows and reproduces.

Big Question

How do some plant structures function?

Plants do not exhibit behaviors in the same way as animals do. However, plants do respond to what is happening in the environments around them. For example, a plant will wilt when it does not get enough water. Leaves on certain kinds of trees change color in the fall, and the leaves may fall off in winter.

Specialized structures of these redwood trees allow them to grow into the world's tallest plants. The trunk of a large redwood must support its heavy weight. Vast networks of roots help to keep these massive trees from falling over.

Plant Structures Support Growth

A plant's structures, such as leaves and roots, have functions that allow plants to meet their needs. Systems of structures in plants' bodies help them

- absorb sunlight and use its energy to make food,
- take in water from the soil,
- move water from the roots to other plant parts, and
- exchange gases with the surrounding air.

The structures of this young plant provide what it needs to grow into an adult.

Different plants have different kinds of structures. For example, some plants have thorns or bristles to protect them from being eaten, such as the sharp spines of cacti. Most animals don't eat cacti because they don't want to get stuck with the spines. The Venus flytrap, on the other hand, looks nothing like a cactus plant. It grows in a very different kind of climate, and it has unique features, including a spiny leaf that closes around insects to digest them. This helps it get some nutrients it needs.

Inside the leaves of plants are structures that absorb energy from sunlight. These structures carry out a chemical process that supports the plant's survival. This process uses the energy of sunlight to make food for the plant. Some plants need a lot of sunlight to survive. Other plants need less light. But all plants have structures that help to use sunlight to stay alive.

Did you know that plants breathe? But they don't breathe like animals. Plants do not have lungs. Instead, they have tiny holes in their leaves. These holes allow air to enter the leaves. Gases in the air that enter the plant help the plants to make food too. As a result of the chemical process, plants release oxygen and water vapor through the holes in their leaves. Plants cannot survive without this exchange of gases with the air around them.

This microscope photo shows the structures inside the cells that capture energy from sunlight to make food.

Plant Structures Support Responses to the Environment

Plants can have a **response** to their environment. A plant may not have enough light or water to grow. Plants cannot move from place to place like animals. Instead, they respond in ways that help them get what they need to grow. Certain kinds of plants can even move their leaves and flowers to get what they need or to protect themselves from harm.

Vocabulary

response, n. a reaction to some factor or condition

Suppose a plant needs more light. Its stem may bend to face the sun. Some plants can turn so that they face the sun as it moves across the sky during a single day. Other plants slowly grow in the direction of available light.

Many plants' roots grow downward because the plant responds to the pull of gravity. This growth sends the plant's roots in the direction where water is likely to be reached.

Even on its side, a seedling can begin to grow sideways, sensing the direction of what is up and what is down.

Plant Growth and Responses Support Survival

Forest fires, floods, insects, and disease can affect a plant's survival. Unlike animals, plants cannot move to a new place to live. Plants, however, have structures and responses that help them survive harmful conditions. In this way, plants can continue to grow and mature.

Look at the picture of the redwood tree. Redwood trees grow in areas where insects, disease, and fire kill many other trees. Yet redwood trees survive many harsh conditions. Like all trees, redwoods have bark. However, the bark of redwood trees helps it to resist insects and disease. As the tree matures, its bark grows thicker. The thick bark on older redwood trees even protects them from fire.

This open section of a redwood tree's bark shows how thick the bark can grow.

Plant Structures Support Reproduction

Plants that survive long enough to mature into adults can reproduce. Remember that when organisms reproduce, they are able to make new individuals of the same type. Plants have male and female structures that aid in reproduction.

Pine trees have cones. Male cones release a material called pollen that plants need to reproduce. Female cones on the same tree receive the pollen. When this happens, a seed forms on the cone. A new pine tree can grow from the seed after it drops to the ground.

Many plants reproduce with flowers. Flowers are structures that often attract pollinators such as insects and birds, which move pollen from one plant to another. In some cases, flower structures expose the pollen to the wind or rain, which helps move it to new areas, where it can grow into new plants.

Look closely. You can see the yellow pollen on the bee's legs that it collected as it moved from flower to flower. Without pollinating bees such as this one, certain kinds of plants would not be able to reproduce.

Human Vision and Hearing

Chapter 5

You have read how structures and functions are always linked together in the lives of plants and animals.

Big Question

How do humans see and hear?

Like all living things, human structures and functions are related. Humans have the same levels of biological organization as other living things. They have cells, tissues, organs, and organ systems.

Two structure-function relationships that help us to survive are the senses of seeing and hearing. When you see, you can't control how your eyes see light. When you hear, you can't control how your ears hear sound. These senses seem simple because you don't need to make them happen. Many processes are taking place inside your eyes and ears. These processes allow you to see and hear.

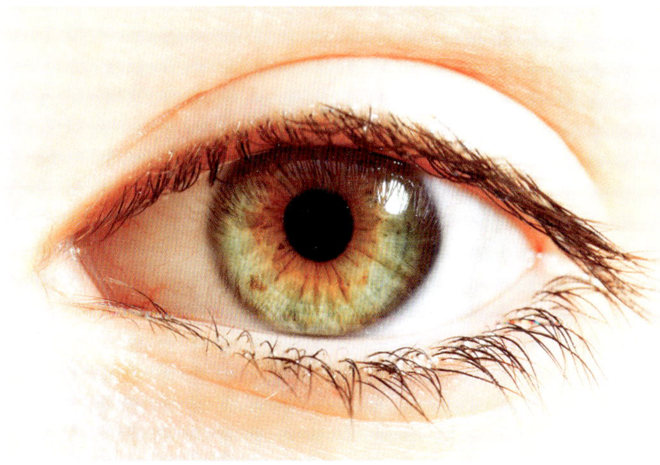

Eyes have many kinds of structures that work together to help a person to see. Look closely at this image of the human eye. Can you see different parts of the eye?

Structures in the Eye Detect Light

The human **visual system** is the set of cells, tissues, and organs that support the function of seeing. Remember that structure is related to function.

> **Vocabulary**
>
> **visual system, n.** the set of cells, tissues, and organs in the eyes and brain that make vision possible

Look at your eye in a mirror. The parts of your eye that you can see are just some of these structures. Other structures inside your eye work as a system to allow vision.

Vision is the ability to see. When you see, your eyes sense light. Light reflects off objects around you. The light waves bounce off an object, and then the light enters your eye through the pupil.

After the light enters your eye, the lens inside the eye bends the waves. The lens is a clear structure like the lens of a magnifying glass. As the light waves bend, they are directed to the rear surface inside of the eye.

At the back inside of the eye, light strikes special cells that transform the light energy into the electrical energy of a nerve signal. These signals move along the optic nerve to the brain. Your brain interprets the signals as an image.

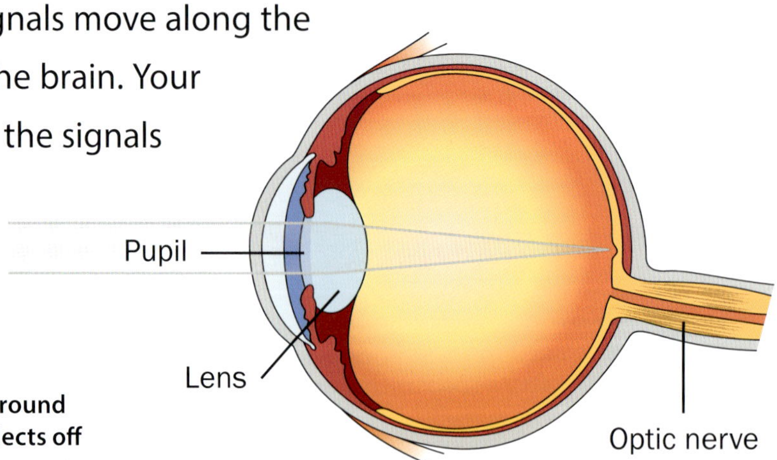

You can see objects around you because light reflects off of them and enters your eye.

Other structures in the rear surface inside the eye help you to see. These structures are made of types of cells. The cells carry out different functions. These include the ability to

- see colors and very small details,
- detect movement on either side of your body, and
- help you see objects when it is dark.

Eyes are helpful for survival. For example, some animals have eyes on the sides of their heads. The placement of these eyes helps the animals to see more of their surroundings. Because of this, they can better spot predators and escape to safety. Other animals have eyes in the front of their heads. This helps them see prey more clearly and to judge distance. In these cases, the eyes help the animals find and capture food. Eyes also help organisms avoid other kinds of dangers, such as obstacles in their path.

These diagrams contrast the fields of vision of an animal with eyes on the sides of its head (left) and an animal with eyes on the front of its head (right).

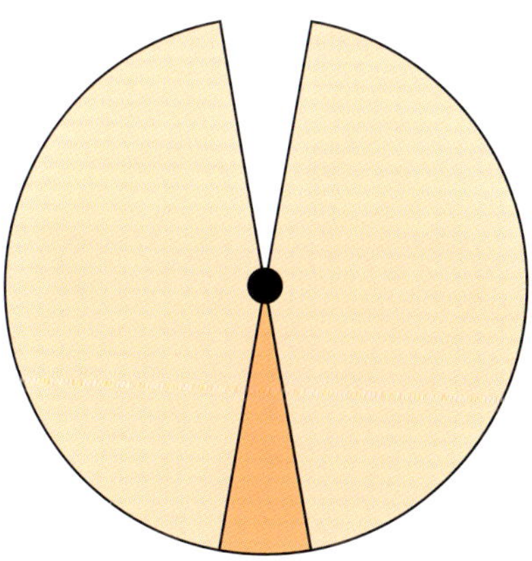

An animal with eyes on the sides of its head has a greater field of vision. It can see objects in more positions around it without needing to turn its head.

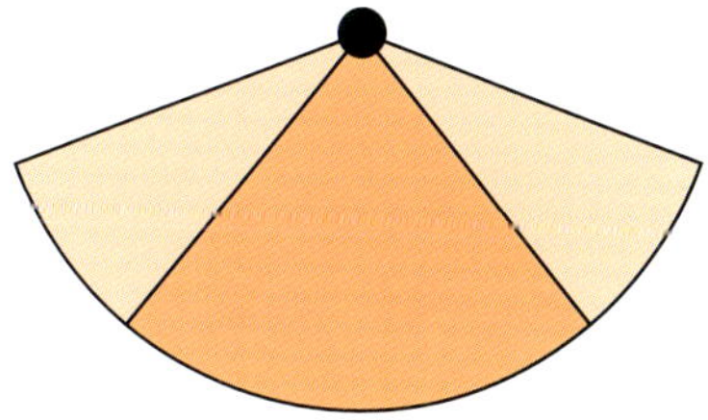

An animal with eyes on the front of its head has a narrower field of vision. But it can generally perceive distance better.

Structures in the Ear Detect Sound Waves

The human **auditory system** is the set of cells, tissues, and organs that allow us to hear. Hearing refers to the ability to detect sound. Sound travels through the air as waves. The structures on the outer ear collect sound waves. The outer ear funnels sound waves toward structures inside the ear. Structures inside the ear convert sound waves into electrical energy that flows through nerves. The brain interprets this nerve signal as sound.

Vocabulary

auditory system, n. the set of cells, tissues, and organs in the ears and brain that enable hearing

Let's take a closer look at this process, remembering that structure and function work together.

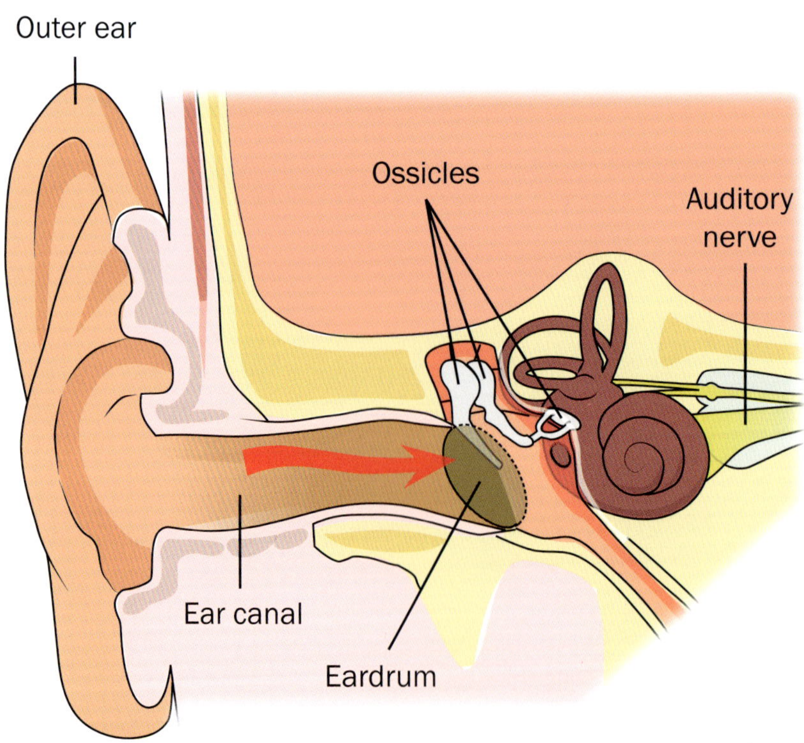

The structures of the outer and inner ear also

- help you determine where sounds are coming from,
- help you figure out how far away sounds are,
- detect high-pitched and low-pitched sounds, and
- help you keep your balance.

Sometimes, structures in the ear do not work the way they should. This can lead to poor hearing. A baby might be born without the ability to hear. Even people with good hearing might lose their ability to hear later in life. We have devices that help people with poor hearing. Hearing aids fit in the outer ear. They make sounds louder and easier to hear. Special implants can carry out the function of damaged structures in the inner ear. The implants help to boost electrical signals to the brain.

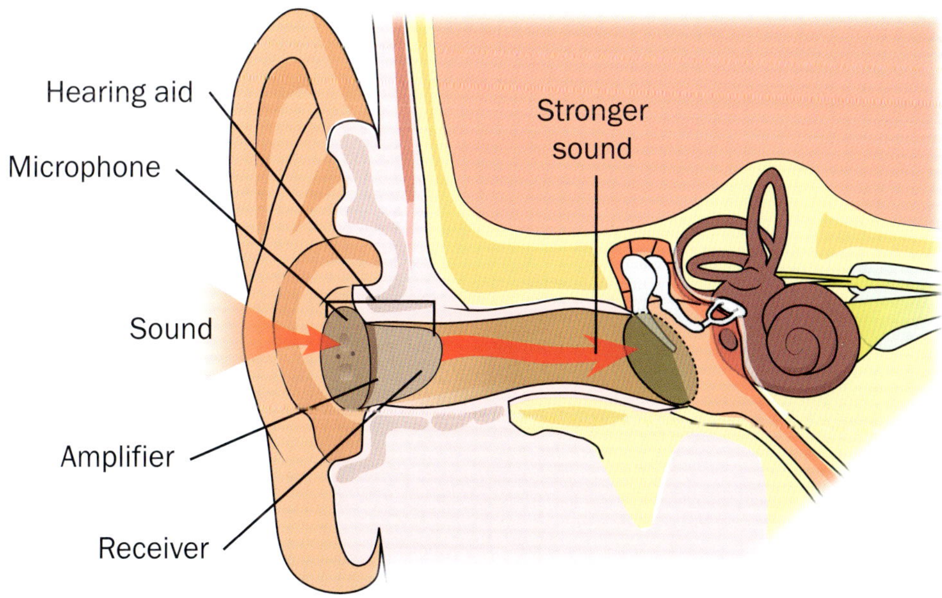

A hearing aid makes sounds louder and stronger.

Hearing and Vision Keep You Safe

Your surroundings are full of sights and sounds. Structures in your visual and auditory systems help you detect conditions around you. These systems send signals to your brain. The signals travel through structures in the nervous system. The brain interprets these signals, which you know as sights and sounds.

Being able to detect sights and sounds helps to keep you safe. Suppose you're on the way to school, waiting to cross the street. You can see the crosswalk sign. Your brain can then determine when it is safe to cross. During a storm, you hear thunder and see lightning. Your brain knows that it is not safe to go outside. You approach a railroad crossing while walking or riding your bike. You hear clanging sounds and see flashing lights. Your brain knows that these are signs warning you to stop and wait.

What sights do you see and sounds do you hear that tell your brain that a train is approaching?

Animal Senses

Chapter 6

Nonhuman animals have structures that allow them to see and hear things that exist around them. The brain processes the information sent from the eyes and ears. An animal uses this information to decide how to respond to the sights and sounds.

Big Question

How do animals sense and respond to their environments?

Sights or sounds can cause an animal to respond. Each is a **stimulus**. A stimulus is an object or event that causes a response.

A stimulus may cause a voluntary response. Think about an animal that **senses** a nearby predator. It may decide to run. But it could decide to hide instead.

Vocabulary

stimulus, n. a factor or condition that causes a response

sense, v. to detect a stimulus

Animals have many responses that are involuntary. This is a behavior that is automatic and done without thought. For example, if something is moving quickly toward an animal's eye, the animal may close its eye or lower its head in response without even thinking about it.

In the winter, many cats' fur grows thicker. Shorter days signal the coming of winter. This is the stimulus. The thicker hair is a response to help keep the animal warm.

Response to Stimulus Varies Among Animals: Sight

Have you ever heard the term *eagle eye*? It is used to describe someone who can spot things easily. Eagles have amazing eyesight compared to most other animals. Their eyes are large. They can also look forward and sideways at the same time. Eagles can see prey, such as a rabbit, from up to three miles away! Owls are another type of bird with large eyes. These eyes help them see to capture prey at night.

Some animals have very poor eyesight. They must rely on other senses to sense their surroundings. Bats hunt at night. They have tiny eyes, so they cannot see their prey well. Instead, they send out a series of shrill squeals. These sounds bounce off objects. When the sound returns to the bats' ears, the bats interpret the sound. In this way, the bat can tell the size and distance of an object. It can also tell whether an object is moving or still.

An eagle may weigh only ten pounds, but its eyes are about the same size as human eyes.

Response to Stimulus Varies Among Animals: Sound

Most animals hear like you do. They detect sound waves. Like you, they have ears to listen out for sounds. Consider elephants. They have very large ears to funnel and hear sounds.

Some animals can detect sounds that we can't hear. The greater wax moth can hear sounds nearly fifteen times higher-pitched than sounds humans can hear. With such sensitive hearing, the moth is better able to avoid predators. The moth's main predators are bats. A bat's call is very high-pitched. This sound is beyond the hearing of most animals. But the greater wax moth can hear the bat's call. The sound acts as a warning to the moth.

Look at the large size of the elephant's ears. An elephant's ears help capture and funnel sounds.

Response to Stimulus Varies Among Animals: Smell

Bears have one of the keenest senses of smell of all land animals. The area inside the bear's nose is up to 100 times larger than a human's. Because of this, bears can smell prey from miles away. Because of this keen sense of smell, bears are able to locate food even though it's far away.

Snakes are unusual because they smell with their tongues. Look at a snake moving. You can see it flick out its tongue. The tongue picks up scent particles nearby. The snake uses organs inside its mouth to detect the scents. Snakes have poor eyesight and hearing. Instead, they rely on their sense of smell. Their sense of smell is how they find food and avoid predators.

A yellow-bellied water snake's tongue is forked so that it can smell more at once.

Response to Stimulus Varies Among Animals: Touch

Without the sense of touch, animals would not be able to feel pain. Without being able to feel pain, they would get hurt more easily. Why? It is because pain is a warning signal from the environment.

Pain is one result of the sense of touch. Some animals have such a strong sense of touch that they can detect vibrations. For example, spiders have special structures on their legs and bodies. These tiny slits and hairlike spines pick up vibrations on the spider's web or other surfaces.

A spider responds when its web vibrates. A spider uses this sense to know when its web has caught prey. The spider can even figure out the kind of prey. The type of vibration can also tell the spider if a predator looking for food finds its web. Humans do not have nearly the sense of touch that spiders have.

Can you see the hairlike structures on the spider's body? They enable the spider to detect vibrations.

Response to Stimulus Varies Among Animals: Taste

Most animals can taste things in their environment. This helps them to know which foods to eat or not to eat to help them stay safe and healthy. Humans have taste buds on their tongue. Other animals have taste buds in different parts of their bodies.

The animal with the strongest sense of taste may be the catfish. While humans may have 10,000 taste buds, catfish can have up to 175,000, depending on how big they get. What enables them to have so many taste buds? The taste buds do not appear in the catfish's mouth but rather all over its body. Most of those buds, however, are located in the catfish's whiskers, which are near its mouth.

Because of the number of taste buds they have, catfish can use their sense of taste to find prey. Catfish are bottom dwellers, which means that they usually live on the floors of rivers, ponds, and lakes. This area is often very dark. While catfish do have eyes, they use their sense of taste to locate food.

Look at the catfish's whiskers. These are where most of its taste buds are located. Because of them, it can detect food in the water even when there is little light.

Plant Responses

Chapter 7

If you grow a plant by a window, it may lean toward the light. This is one example of how plants respond to a stimulus. The stimulus is sunlight, and the response is that the plant bends.

Like animals, plants can sense what is happening in their environment. However, plants do not have the same structures as animals. Plants do have responses that help them get energy and water and nutrients. We call these responses **tropisms**. The word comes from ancient Greek and means a turning.

Big Question

How do plants sense and respond to their environments?

Vocabulary

tropism, n. a plant's growth or movement in response to a stimulus

A well-known plant tropism is the response to light. How do plants move so that they can face toward light? Cell enlargement on one side of a stem causes the plant to lean. As the plant leans toward light, its leaves can absorb more of the light.

Plants will often grow toward the sunlight to better meet their energy needs.

Plants Respond to Gravity

Plants also respond to gravity. A plant's parts respond to Earth's gravity. Some parts grow down, toward the pull of gravity. Others grow up, or away from the pull. The growth of a plant part away from or toward the pull of gravity is another kind of tropism.

Suppose you were to plant a seed sideways. In which direction do you think the roots and shoots would grow? If you think straight out, you are right—at first! But soon the root would start to grow downward. The shoot would turn upward. Structures in each plant cell signal them to grow in these directions.

A plant's responses to light and gravity are tropisms that help it to meet its needs. A plant's roots grow down, away from light and toward the pull of gravity. The roots grow this way so that the plant can reach the water and nutrients it needs.

How are the shoot and roots from this seed responding to light and gravity?

Plants Respond to Temperature

In fall, leaves change color. This change is the trees' response to shorter days and lower temperatures. In fall, the weather becomes cooler, and there is less daylight. Leaves do not make as much food. The chemical that gives leaves their green color begins to break down. Other chemicals produce red, yellow, and orange colors in the leaves. Later in fall, trees drop their leaves. They store the energy. The trees also save water.

Plants also respond to high temperatures. A plant's stem, leaves, and flowers may droop or wilt. Wilting happens when a plant loses more water (in the form of water vapor) than its roots take in. In hot, dry weather, the soil may have too little water to meet the plant's needs.

These are not examples of tropisms. Why? The plant is not growing or turning its direction of growth.

Changes in daylight and temperature cause a tree's leaves to change color.

Plants Respond to Touch

The Venus flytrap is a carnivorous plant. It gets some of its nutrients from insects. How can this plant catch insects? It doesn't have eyes or ears. Instead, the Venus flytrap responds to an insect's body weight using a sense of touch.

Venus flytraps have tiny hairs that act as sensors. When a hair is touched twice in a short amount of time, the plant's leaves snap shut. The insect becomes trapped inside the leaves. Here, the plant digests the insect. Then the leaves can absorb the nutrients the plant needs.

Plants will also respond to contact with objects. For example, when a plant's roots encounter a rock in the soil, they will grow around it. Some plants, such as pea plants, grow vines or tendrils. The vines support the plants as they grow. The vines wrap around nearby objects, allowing the plant to grow taller than it could otherwise. These are both examples of tropisms.

When a plant grows toward light, the movement happens too slowly for you to see it happening. The Venus flytrap moves quickly enough that you can see it when it closes to trap an insect.

Helen Keller and Louis Braille

Chapter 8

We have a visual system and an auditory system to help us survive. These systems sometimes do not function correctly. Some people have poor vision and/or hearing. Others may have total deafness or blindness. People impaired in these ways must rely on other ways to monitor the environment. These are used instead of sight or sound so that such people can sense the world around them.

Big Question

What have I learned about structures and functions?

Words to Know

Deafness means the complete loss of hearing. *Blindness* means the complete loss of vision.

Helen Keller

Helen Keller was born in Alabama in 1880. At birth, she was healthy. But at the age of 19 months, she became very ill. Due to the illness, she became deaf and blind. She also could not speak well enough to be understood by others. She had to learn other ways to monitor her environment and to communicate.

Helen Keller lost her ability to hear and see when she was very young.

At that time, no adaptive devices could help Helen hear or see. Instead, she used her hands. Our fingertips are full of nerve endings. Hence, they are very sensitive. Helen relied on her sense of touch. Through touch, she "saw" the world around her. She also used touch to communicate with others.

Word to Know

Adaptive means able to adapt. An adaptive device is an object that has been "adapted" to help with a daily task. Hearing aids, eyeglasses, and wheelchairs are examples of adaptive devices. These are conceived by engineering designers and built by engineers.

Helen had a teacher, Anne Sullivan. Anne used sign language to teach Helen letters and words. First, Anne would spell the name of an object. She did this by writing each letter onto Helen's hand. Then she had Helen hold and touch the object. Soon Helen was able to use sign language to communicate with others.

Helen also learned to use touch to "read" lips. When Anne talked, Helen touched her mouth to feel the way it moved. Later, she was even able to read books. These books were printed with a special alphabet. This alphabet has a pattern of raised dots to represent each letter.

Helen and her teacher, Anne Sullivan

Louis Braille

Louis Braille developed the special alphabet that Helen Keller used to read. Louis Braille was born in France in 1809. Like Helen, he was born with good vision. But when he was three years old, he lost his sight after an accident.

Louis Braille

Louis went to school like other children. However, he was frustrated that the only way he could learn was by listening. He wanted to be able to read, too. Louis heard about a code that soldiers used to communicate at night during battles. This code was made of raised dots on paper. He studied the code. He then used it as the basis for his new alphabet. By the time he was fifteen, Louis had developed a new system of raised dots. The system, named after him, is known today as *braille*. The first book written in braille was printed in 1829.

Blind people and those with impaired vision can use their fingers to read braille. They touch each letter to feel the pattern of dots. The letters and words are arranged as in a regular book. Today, braille is used in many places other than books. You can find it used in many public places. People can use a special printer to print out braille documents from a computer.

Braille and sign language help blind and deaf people lead normal lives. They show that humans can adapt when senses don't work as they should. Helen Keller and Louis Braille were able to adapt. In this way, they conquered the hurdles they had as children.

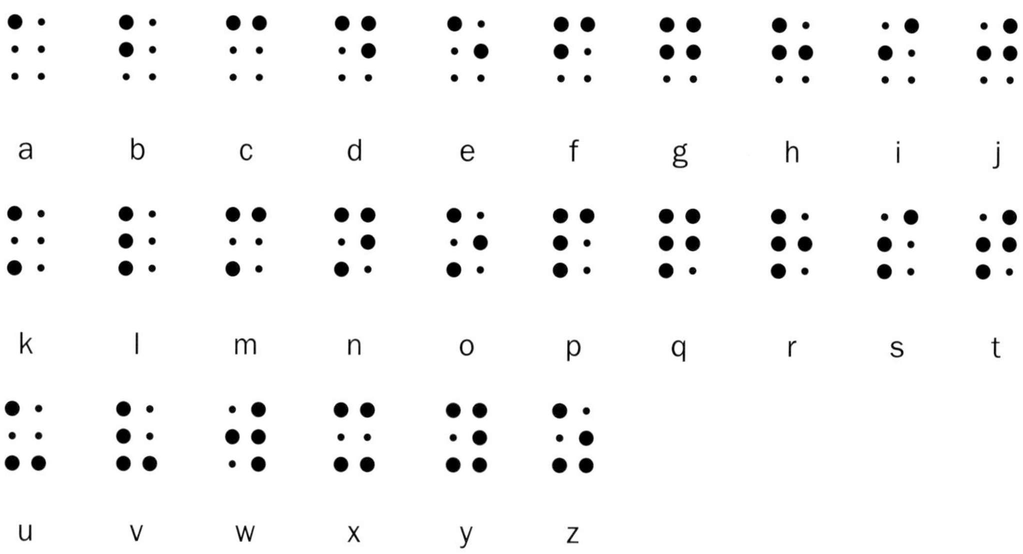

Each braille character contains up to six dots in different patterns.

Glossary

A

auditory system, n. the set of cells, tissues, and organs in the ears and brain that enable hearing (24)

B

behavior, n. the ways in which an organism acts (13)

C

cell, n. the smallest unit of life (1)

F

function, n. the way that something works to achieve a task or serve a purpose (5)

G

growth, n. the life process of becoming bigger and stronger (12)

N

nerve, n. a bundle of cells that send and receive messages between the brain and other body parts (10)

O

organ, n. a body part made up of related tissues that perform a specific function (6)

organ system, n. a group of organs that work together to carry out certain functions (8)

R

reproduce, v. to make new, young individuals of the same type (14)

response, n. a reaction to some factor or condition (18)

S

sense, v. to detect a stimulus (27)

stimulus, n. a factor or condition that causes a response (27)

structure, n. the arrangement of parts that make up something (5)

survival, n. the state of continuing to live (11)

survive, v. to stay alive (11)

T

tissue, n. a group of similar cells joined together (4)

tropism, n. a plant's growth or movement in response to a stimulus (33)

V

visual system, n. the set of cells, tissues, and organs in the eyes and brain that make vision possible (22)

Core Knowledge®

CKSci™

Core Knowledge SCIENCE™

Editorial Directors

Daniel H. Franck and Richard B. Talbot

Subject Matter Expert

Joyce Latimer, PhD
Professor
Department of Horticulture
Virginia Tech University
Blacksburg, VA

Illustrations and Photo Credits

Andrey Armyagov / Alamy Stock Photo: 5
anton havelaar / Alamy Stock Photo: 26
Benny Marty / Alamy Stock Photo: 11
buccaneer / Alamy Stock Photo: 3
Caia Images / SuperStock: 21
Chronicle / Alamy Stock Photo: 39
Cultura Limited / Cultura Limited / SuperStock: 29
Everett Collection Inc / Alamy Stock Photo: 38
Fero Bednar / Alamy Stock Photo: i, iii, 36
FLPA / SuperStock: 34
Gallo Images / SuperStock: 13
Grant Heilman Photography / Alamy Stock Photo: 33
Harry Collins / Alamy Stock Photo: 28
imageBROKER / Alamy Stock Photo: 35
imageBROKER / SuperStock: 14, 27
J M Barres / age fotostock / SuperStock: 4
Jack Bailey / Pantheon / SuperStock: 12
Jef Meul / Minden Pictures / SuperStock: 20
Library of Congress / SuperStock: 37
Martin Shields / Alamy Stock Photo: 18
Matthijs Kuijpers / Alamy Stock Photo: Cover E, 30
Morley Read / Alamy Stock Photo: 31
Pete Niesen / Alamy Stock Photo: 19
Prisma / SuperStock: 32
robertharding / SuperStock: Cover A, 17
Science Photo Library / SuperStock: 1, 2
Scott Wilson / Alamy Stock Photo: 15
Valery Yurasov / Alamy Stock Photo: 16
Zoonar GmbH / Alamy Stock Photo: Cover B, 10a